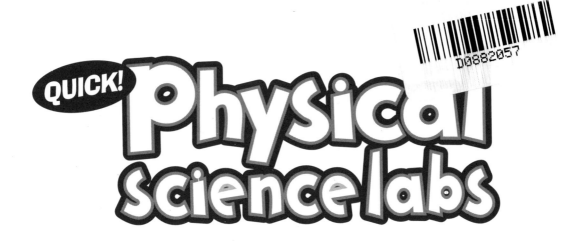

QUICK! Physical Science labs

Deborah C. Gutman
Jennifer L. Boudart
Carol A. Thornton

Illustrated and Designed by: Cox & Co.
Cover Illustrations by: Lopata Design
Research Assistance by: Katie Davis and Marilyn Sams

ISBN: 1-56911-805-1

Printed in China.

Table of Contents

Introduction

"Learning science is something students do. Full inquiry involves asking a simple question, completing an investigation, answering the question, and presenting the results to others." National Science Education Standards

Quick Physical Science Labs provides active, hands-on science investigations in the content strands of physical science for students in grades K-2. Each unit in this book ties in with a specific program standard outlined in the National Science Education Standards. These connections are highlighted at the beginning of each unit. Investigations address fundamental physical science concepts including properties and states of matter, energy, and forces.

Stimulated by the challenge of simple questions, children:
- complete investigations or other organized activities such as structured games
- organize data or gather information to answer questions
- present results to others either orally or in writing
- apply what they learn to other areas

In the spirit of the National Science Education Standards, activities require students to use scientific inquiry to investigate a question, communicate findings, and connect learning to real-world experiences. Students are challenged to use scientific tools and to work cooperatively when completing activities and sharing results.

Each unit investigation includes Teacher's Notes and these reproducible student pages:

Quick Lab Experiment—introductory activity

Data Collection/Communication—record sheet for Quick Lab findings

Working Together—cooperative group project

Using Tools—research using a scientific tool to gain information

Learning Link—interdisciplinary connection

You can reproduce student pages or use them as transparencies. Consider having students create a book to hold notes and drawings relating to the activities. Check the equipment table at the back of this book for materials needed in each unit as well. As children engage in the investigations, activities, and games, they will learn the essence of scientific inquiry—a refinement of youngsters' natural curiosity to learn and understand the world around them!

Water

Materials can exist in different states: solid, liquid, and gas. Some common materials, such as water, can be changed from one state to another by heating or cooling.

National Science Education Standards

Getting Started

This unit introduces students to properties of water and water in its different states. Gather information online by entering the query: "water cycle" AND "teaching" AND "science." Share books such as **A Drop Of Water** (Walter Wick, 1997) or **Water** (Graham Peacock, 1994). Share the background information below. Check the Equipment Table for materials needed to complete the activities in this unit.

About Water

- Water is made of two elements: hydrogen and oxygen (H_2O).
- Water naturally exists in three states: solid (ice), liquid (water), and an invisible gas (water vapor).
- When water freezes into ice, it expands, or takes up more space. It also becomes less dense. This is why ice floats in water.
- Water covers three-fourths of the Earth's surface; two-thirds of the human body is water.
- The earth's water supply is constantly recycled via the water cycle. Water falls from clouds as liquid or solid precipitation. It collects on the Earth's surface in ponds, lakes, oceans, rivers and other bodies. It also collects underground. Water returns to the air as it evaporates from Earth's surface due to warming, turning into water vapor. As the air cools, water vapor condenses and falls once more.
- Many substances dissolve easily in water.
- Water has buoyancy. It supports objects that are less dense than it is. An object's ability to float or sink cannot always be predicted by its size or shape or what it contains.

Misconceptions

Students would likely guess that solid ice is heavier (more dense) than liquid water. They might also believe solid ice takes up less space than liquid water, which is not the case.

Teaching The Unit

Quick Lab and Data Collection/Communication: Begin the activity with a discussion of things students have seen floating on water or sinking beneath it. Ask students to explain why they think an object sinks or floats. Tell them they can test these explanations by trying to predict which objects in the Quick Lab will sink or float. For Think About It on the Quick Lab, you might get students started by suggesting they turn their foil square into a toy boat that can hold something that might otherwise sink. Discuss how students' thinking has changed by reviewing Think About It on the Data Collection page.

Working Together: Introduce the activity by reading books or poems about rain and clouds, such as **The Little Cloud** (Eric Carle, 1996), **Sheep on a Ship** (Nancy Shaw, 1989), or **Cloudy with a Chance of Meatballs** (Judi Barrett, 1978). Review the water cycle and associated vocabulary. Discuss how water may change form during each stage. Explain that water changes form, or state, when it heats and cools. Share examples of water turning into steam or water freezing into ice. Model the water cycle using a coffeemaker filled with hot water: Remove the pot's cover and place a plate of ice cubes sprinkled with salt (to speed melting) on top of the pot. Watch and wait for a wispy "cloud" to form. Lift the plate to reveal the "raindrops" that form on the bottom. Students should label their drawings as follows: 1. evaporation 2. water on Earth 3. precipitation 4. vapor in clouds. For Think About It, students should answer that water can be a solid, liquid, or gas and it changes form as its temperature changes.

Using Tools: Discuss evaporation and ask students where they have seen evaporation at work (a puddle drying in the sun, boiling water, a dryer heating clothes). For Think About It, students should answer that liquid water turned into water vapor and that the sun's heat helped this happen.

Learning Link: This activity introduces the concept of water's ability to mix with other substances. After observing the behavior of oil in water, help students research oil spills for Think About It, and consider testing water's ability to mix with or dissolve other materials.

Sink or Float?

PREDICT Which of these things will sink? Which ones will float?

 Write   an S or an F in each box: S = sink   F = float

Penny

Soap

Plastic Cup

Foil Square

Key

Marshmallow

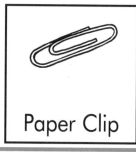

Paper Clip

Feather

LAB TIME

Drop each object in ⌇⌇⌇ .

water

👁 what happens.

Watch

THINK ABOUT IT

Can you make a floating object sink or a sinking object float?

Try changing its shape.

Try adding it to ⌇⌇⌇ with another object.

Tell or ✎ about it.

write

Sink or Float?
Fill out this page as part of your lab.

LAB TIME

Cut and glue pictures to show which objects sink or float.

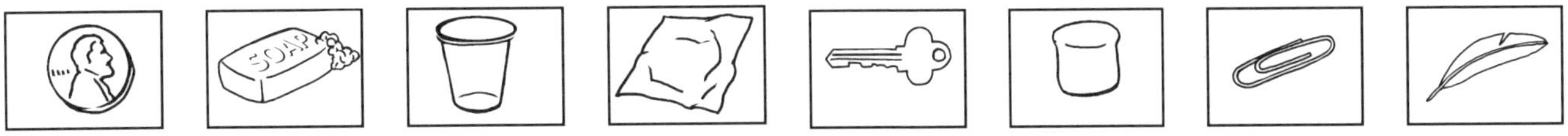

SINK

FLOAT

THINK ABOUT IT

How well did you predict which objects would sink or float? Can you tell if something will sink or float just by looking at it? Tell or ✏ about it.

a number sentence: _____ + _____ = _____

Floating Objects

Sinking Objects

Total

Watch a Water Cycle

PREDICT

Talk about the water cycle with your teacher.
How does 〰〰 change as it travels the water cycle?

LAB TIME

👁 your teacher make a water cycle indoors!

1. 👁 for ways this water cycle is like a real water cycle.

2. 👁 for 〰〰 in different forms.

3. ✏️ these pictures in the correct places.

| water vapor in clouds | water on Earth | water precipitation | water evaporation |

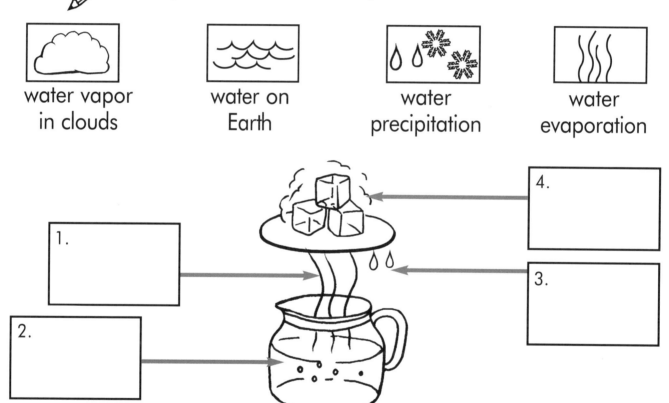

THINK ABOUT IT

Finish this sentence: 〰〰 can have 3 forms:

_____ _____ _____

1. _____ 2. _____ 3. _____

When 〰〰 is on Earth, how do we use it? Tell or ✏️ about it.

Measuring Cup

Use a 🥛 to 👁 〰 "disappear."

measuring cup

〰 can "disappear!" 〰 turns into a gas you cannot see. The gas is called **water vapor**. Water vapor is in the air all around you.

LAB TIME

1. Fill a 🥛 with 〰 to the 8-ounce mark.

2. Place the 🥛 where ☀ shines on it.

sun

3. Measure how much 〰 is in the 🥛 each day. ✏ your measurements below. Do this until no 〰 is left in the 🥛.

Day	Water (oz.)

THINK ABOUT IT

Finish these sentences. _____

The 〰 turned into _____ .

_____

_____ from the ☀ helped the 〰 disappear.

Water

Drip-Drop

〰〰 is good at mixing with other things. What things can you think of? Some things do not mix well with 〰〰. Mix 🍾 with 〰〰 and 👁 what happens.

1. Use an 💧 to drip 🍾 on waxed paper.
   eyedropper

   Drip 〰〰 on the waxed paper, too.

2. Look at the shape of each drop from the side. Draw ✏ each drop's shape.

Drop

〰〰 **Drop**

3. Drip a few drops of 🍾 into a cup of 〰〰. ✏ a picture.

4. Stir the 〰〰 with a toothpick. ✏ a picture.

THINK ABOUT IT

If ships spill 🍾 in 〰〰, it does not go away easily. What do you think happens to 〰〰 plants and animals? Tell or ✏ about it.

Water

Magnets

Magnets attract and repel each other and certain kinds of materials.
National Science Education Standards

Getting Started

Introduce this unit by sharing books such as **First Step Science — My Magnet** (Robert Pressling, 1994) or **A New True Book: Experiments with Magnets** (Helen J. Challand, 1986). Gather information online by entering the query "magnets" AND "science projects" AND "learning." Also share information listed below. Check the Equipment Table for materials needed to complete the activities in this unit.

About Magnets

- Magnetism is defined as the ability to attract iron.
- Greeks discovered lodestones (iron magnetized by lightning) about 2000 years ago. The Chinese made the first compasses the 12th Century.
- Every magnet has two poles: north and south. If you break a magnet, the two resulting magnets will each have two poles.
- Opposite poles (north-south) attract while like poles (north-north or south-south) repel.
- A magnetic field is produced by the motion of electrons in the iron atoms of the magnet.
- The Earth is a huge magnet! Electric currents deep in the Earth may create the Earth's magnetic field. The Earth reverses its magnetic poles on an irregular schedule.
- Even atoms are tiny magnets!
- Magnetized material coats a computer disk, allowing information to be stored. Magnets can de-magnetize or erase computer disks, videotapes, and credit card codes.

Misconception

Children may believe all metals are magnetic. This is not correct. For example, a magnet will not pick up an alluminum soda can. Some metals (iron, nickel, cobalt) attract, while others (aluminum, copper) do not.

Teaching the Unit

Quick Lab and Data Collection/Communication

Define magnetism for the class. Help students select ten objects to test for magnetism. Be sure to include several magnetic metal objects. Review the graphic organizer with students to help them summarize and analyze their results. For Think About It, help students test several metal objects that a magnet will not attract.

Working Together

All magnets have two poles. Like poles repel; opposite poles attract. Discuss this idea with students. You will need to help students identify the North and South poles on their magnets to test what happens when like and opposite poles are placed together. To do so, suspend a bar magnet by tying a piece of string around its middle. When it stops swinging, use a compass to identify its north-seeking pole. The opposite pole is south-seeking. Label the poles on the bar magnet. Use the bar magnet to identify north and south poles on the other magnets by using the fact that likes repel and opposites attract. Tell students to draw the results of their tests. They should indicate that north-north and south-south repel, while north-south poles attract.

Using Tools

Practice using the balance and cubes with familiar objects first. Use 3 strong magnets that can hold many clips. Based on results, help students fill in the scale for the number of paper clips on the left side of the graph. Students will discover that size or shape do not necessarily indicate magnet strength. As an extension, use a mass set to find the mass in grams of the paper clips as an extension activity.

Learning Links

Students can design paper or even cloth racing paths for the turtles. A magnet held under the track will attract the paper clip on the turtle and pull it along. If students attach a magnet to the turtle, holding the like pole of a magnet under the track will push it along, while an opposite pole can be held under the track to pull the turtle along.

Pick Me Up, Magnet

PREDICT

Collect 10 objects to test with a magnet. or the name of
each object in the spaces.

Draw   write

Do you think a magnet can attract each object? "A" for
attract in the box if you do. "NA" for not attract if you do not.

1. -----------------

2. -----------------

3. -----------------

4. -----------------

5. -----------------

6. -----------------

7. -----------------

8. -----------------

9. -----------------

10. -----------------

LAB TIME

Test your objects to see if they are magnetic.
1. Hold a magnet near each object.
2. If the object "sticks" to the magnet, it is magnetic.
   If the object does not stick, it is not magnetic.

THINK ABOUT IT

Did your magnet pull on some objects more strongly than others?
Which ones? Tell or about it.

Magnets

Pick Me Up, Magnet

Fill out this page as part of your lab.

LAB TIME

or ✏ the name of each object in the correct box.

Magnet Attracts (A)	Magnet Does Not Attract (NA)

Total: _____   Total: _____

Sort your objects by what they are made of, and whether or not a magnet attracts them.

	Attracts	Not Attracts
Metal		
Non-Metal		

THINK ABOUT IT

Can your magnet pick up all metal objects? Try a test!

Tell or ✏ about it.

✏ write

Magnets

Pole Power

The ends of a magnet are called poles. One pole is called "north."
One pole is called "south." Each pole has a strong force.

PREDICT

What happens when you bring magnet poles together?

LAB TIME

Find the poles on each magnet. the letters "N" for north
and "S" for south on each magnet below.

Horseshoe   Bar   Circle

Try 3 tests, each with 2 pairs of magnets. what you
the poles doing in the boxes below.

1. N to S   2. N to N   3. S to S

Attract (Stick Together)	Repel (Push Away)

THINK ABOUT IT

Finish these sentences.

Magnets have _____ poles.
Poles attract when they are **different/the same.** (circle one)
They repel when they are **different/the same.** (circle one)

Magnets

Balance Scale and Centimeter Cubes

Use a ⚖ and ▱ to measure magnet strength.

balance scale   centimeter cubes

LAB TIME

Practice massing objects with a ⚖ and ▱. Practice making the ⚖ level.

Choose 3 magnets to test for strength.

1. Pick up as many 📎 from a pile as you can with one magnet.

paper clips

2. Place the 📎 on one side of the ⚖.

3. Add ▱ to the other side of the ⚖. Make the ⚖ level.

4. Fill in the graph. Fill in the scale on the left side. Your teacher will help you. Color the boxes to show how many ▱ you added for each magnet.

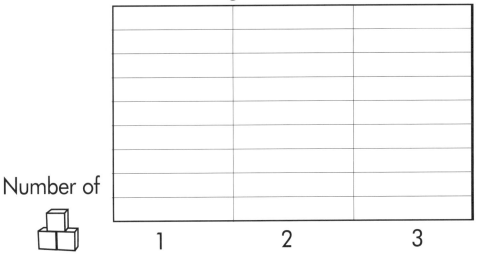

Number of

▱   1   2   3

THINK ABOUT IT

Which magnet is strongest or weakest? How can you tell?

Tell or ✎ about it.

Magnets

Magnets in Motion

Some magnets can attract objects even when another material is between them. Use this fact to make a racing game!

Use magnets to make a 🐢 race around a 🏁.

turtle   race track

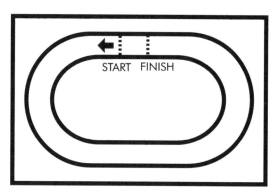

1. ✂️ Cut and color a 🏁 on a large piece of paper.

   It should look like the one above.

2. ✂️ and color your 🐢.

3. Tape a 📎 or small magnet to the 🐢.

   Place the 🐢 on the 🏁. Hold a magnet under the 🏁

   so it attracts the 🐢. Make the 🐢 race!

THINK ABOUT IT

How does the 🐢 move? Can you design a way to race against your friends? Try it!

Light and Heat

Light travels in a straight line until it strikes an object. Light can be reflected by a mirror, refracted by a lens, or absorbed by the object. Heat can be produced in many ways such as burning, rubbing or mixing substances. Heat can move from one object to another by conduction.

National Science Education Standards

Getting Started

To gather information online, enter the query "property of light" AND "teaching" AND "children." Also share books such as **Sound and Light** (David Glover, 1993) or **What Makes a Shadow**? (Clyde Bulla, 1994), as well as background information listed below. Check the Equipment Table for materials needed to complete the activities in this unit.

About Heat and Light

• Light and heat both are part of the electromagnetic spectrum generated by the sun's radiation. Light rays are visible. Heat is made of infrared rays, which can be felt but not seen by the naked eye. A fire can provide both light and heat.

• Light passes through transparent objects (glass), is partially blocked by transparent objects (e.g., paper), and is completely blocked by opaque objects (e.g., wood).

• Sunlight is made of a spectrum of colors: red, orange, yellow, green, blue, indigo, violet. When light hits an object, the color(s) it reflects are the color(s) we see. For example, grass reflects mostly green light.

• Temperature is a measure of an object's heat energy. If two objects touch, heat passes from the warmer object to the cooler object until equilibrium of temperature results. Temperature equilibrium is reached via conduction in solids, convection in liquids and gases, and radiation through empty space.

• Light travels very quickly in a straight path. For example, light from the sun takes about eight minutes to reach you! When light hits an object it may be reflected (bounced back), refracted (bent) or absorbed. Shiny surfaces are good reflectors. Water is a good refractor. Black paper is a good absorber.

Misconception

Students may not realize that a rainbow does not have seven distinct bands of color. In reality, the bands blend together where they meet, creating additional color combinations that are not readily seen.

Teaching the Unit

Quick Lab Experiment and Data Collection/Communication: A bright day provides enough sunlight for this activity. Have students set up their experiments near a window. Students may need help tilting their mirrors to create a spectrum on the white card. Students should see a rainbow spectrum of color: red, orange, yellow, green, blue, indigo, and violet. For Think About It, students should see that the milky water blocks some of the colors (only the orange and red wavelengths are seen). Discuss rainbows with children, including the fact that only half a rainbow is visible above the horizon.

Working Together: Discuss the terms reflection, refraction, and absorption. Help students "catch" the flashlight beam on the white card. Students' drawings and answers for Think About It should indicate that the black paper absorbs (soaks up) the light, the mirror reflects (bounces back) the light, and the water bends the light in a new direction.

Using Tools: Help students practice reading a thermometer. Discuss the scale (Celsius or Fahrenheit), and remind students to ndicate the scale in their measurements. For Think About It, discuss how people use thermometers, e.g., to measure body temperature, outdoor temperature, cooking temperature, and water temperature.

Learning Link: Use a magnifying lens to observe an old TV screen and demonstrate the introductory statement. When students spin their tops, they will see the red, green, and blue wedges "mix" to create rings of new colors. For Think About It, students should find that blue + yellow = green, red + blue = purple, and red + yellow = orange.

Quick Lab Experiment

Rainbow Colors

Light from the ☀ looks white. It is really a mixture of many colors.
sun

When light from the ☀ hits ≋ drops, it sometimes shows its
water

seven main colors. These colors form a 🌈 pattern.
rainbow

PREDICT

How does light make a 🌈 when it shines on ≋ ?

LAB TIME

Make a 🌈 indoors! Work with a partner. Use light
from the ☀ or from a 🔦 .
flashlight

1. Pour ≋ into a large jar or pan.

2. Hold a ▱ under the ≋. Tilt the ▱ up.
mirror

3. Let light shine on the ▱ .

4. Hold a white card above the ▱ . Tilt the ▱ to bounce light
on the card. 👁 for colors on the card.
Watch

THINK ABOUT IT

Do you think colored ≋ makes 🌈 too? Find out! Let light
shine through milky ≋. Tell or ✏ about what happens.
write

© Learning Resources, Inc.

Rainbow Colors
Fill out this page as part of your lab.

LAB TIME

Which colors make up light from the ☀?  the colors that show on your white card.

Draw

Use the Word Bank to label each color you see.
Be sure to place the colors in order.

Word Bank
green
red
blue
yellow
indigo
violet
orange

THINK ABOUT IT

How was the color pattern like a 🌈?

When do you 👁 for a 🌈 outdoors?

A 🌈 is really a whole circle shape. Why do we see only half a circle? Tell or ✏ about it.

Here A Light, There A Light

A beam travels in a straight line. If the beam hits an object, it might pass straight through, bounce off, or "disappear."

PREDICT

What happens when a beam hits a , , or black ?
paper

LAB TIME

1. Shine a flashlight beam on:

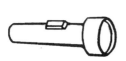

Mirror   Black paper   Glass of water

2. Move a white card until you "catch" the 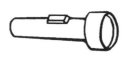 beam each time.

3. 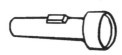 a line to show how the beam travels when it hits each object.

THINK ABOUT IT

Finish these sentences.

The _____ soaked up the light.

The _____ bounced the light back the way it came.

The _____ bent the light in a new direction.

Light and Heat

Thermometer

Use a ✎ to measure temperature.

thermometer

A ✎ is a tool that measures how hot or how cool something is. This measurement is called temperature. The liquid inside a ✎ rises as temperature rises. It falls as temperature falls.

LAB TIME

Practice using a ✎.

1. Start by coloring the warmer arrow RED.
   Color the cooler arrow BLUE.

2. Look at your own ✎. Practice reading the numbers on the side. Your teacher will help you.

3. Use your ✎ to measure these four temperatures.
   Write the temperature in each box.

On my 🪑 [ ]°
desk

after 1 minute in the ☀ [ ]°

after 5 minutes in the ☀ [ ]°

after 10 minutes in the ☀ [ ]°

THINK ABOUT IT

Finish this sentence.

Heat from the ☀ made the temperature _____ .

Can you think of four ways you might use a ✎?
Tell or ✏ about it.

Colors Go Round

Did you know a color TV picture is made with only three colors of light? When red, green, and blue light mix, your eyes see other colors.

Make and spin a top to see the colors mix!

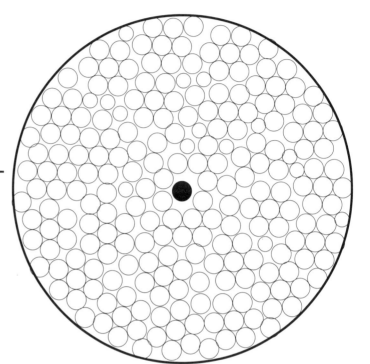

1. Color in the dots. Use red, green, and blue markers. Spread each color throughout the circle. (about 76 dots of each color.)

2. ✂ out the circle. Gently push a pencil through the center.

3. Place one end of the pencil on a 🪑. Hold the top between your 🖐.

4. Spin the top by rubbing your 🖐 together. What happens to the colors?

THINK ABOUT IT

What happens when you mix red, blue, and yellow paints? Find out! ✏ what <u>you see.</u>

Blue + Yellow = _____

Red + Blue = _____

Red + Yellow = _____

Light and Heat

Motion

The position of an object can be described by locating it relative to another object or the background. An object's motion can be described by tracing and measuring its position over time. The position and motion of objects can be changed by pushing or pulling. The size of the change is related to the strength of the push or pull.   *National Science Education Standards*

Getting Started
To gather information online, enter the query "physics" AND "classroom" AND "science." Share books such as **The Magic School Bus Plays Ball – A Book About Forces** (Joanna Cole and Bruce Degen, 1997) or **Force, Of Course!** (M. Leontovich, 1995). Also share background information below. Check the Equipment Table for materials needed to complete the activities in this unit.

About Matter
- An object moves or stops due to a push or pull (force) acting on it. Inertia describes objects at rest or moving at a constant rate in the same direction until forced to speed up, slow down, or change direction. Average speed is a rate of the distance covered in a certain amount of time. Velocity describes both the speed and direction. To calculate average speed, divide distance traveled by the time.
- Gravity is the main force that causes a car to be pulled down a ramp to the ground. Two cars starting at the same height should hit the floor at the same time.
- Friction is a force that resists motion between two objects. Air friction affects falling objects less than surface friction affects sliding or rolling objects. Two objects falling from the same height will hit the ground at the same time, if you do not count air resistance.
- Every object in the universe pulls on other objects with a gravitational force. The mass of the object determines the strength of the pull.

Misconception
Students may think that if two cars roll down a ramp from the same height, the heavier one will reach the bottom first. Neglecting air friction, they reach the bottom simultaneously due to gravity.

Teaching the Unit

Quick Lab and Data Collection/Communication
This activity introduces the concept of speed in a qualitative context. Students will be able to observe from the car's stopping points that the higher the ramp, the greater the speed and the farther the distance traveled. Help students set up each ramp height, and be sure they understand how to mark each car's stopping point. For Think About It, students might experiment with cars of different mass or size, or different ramp surfaces and lengths.

Working Together
Help students mark a 24-foot line of tape on the floor, perhaps in the hallway. Help them mark each foot with tape. Pick music with a slower beat for students to walk to for the activity. Remind students to march to the beat as they walk along the line. You can expand on students' results by linking with math problems, e.g., subtraction number sentences for the distance traveled. For Think About It, students should calculate that they would travel a farther distance if walking to a faster beat and a shorter distance if they took smaller steps.

Using Tools
Help students choose objects to test on their ramps. You might suggest flat-bottomed objects, objects with wheels or rounded bottoms, and objects with textured surfaces. Help students set up the ramp and meter stick. Use students' results to decide which increments to use for the scale (depending on the farthest distance recorded). Answers will vary based on objects chosen. For Think About It, students should recognize that wheels and round bottoms enable objects to move farther than flat bottoms or highly textured bottoms (which create more friction).

Learning Link
Help students practice counting swings. Discuss what might affect the rate of swinging. Students might answer how hard they pump their legs, how high they swing, and so on. Test students' theories if possible. In actuality, the swing acts like a pendulum, e.g., the rate is only affected by the length of the swing chain: the longer the chain, the slower the rate.

Fast, Faster, Fastest

PREDICT

Circle the car that will go fastest.

car

LAB TIME

1. Use a board and books to make a ramp. Roll a toy car from top of the ramp. Mark where it stops with tape.

   board   books   ramp

2. Make the ramp 2 times as high using more books. Roll the car down this ramp. Mark where it stops with tape.

3. Make the ramp 3 times as high using more books. Roll the car down this ramp. Mark where it stops with tape.

THINK ABOUT IT

Can you think of 2 ways to make the car go slower down the ramp? Tell or write about it.

write

Motion

Fast, Faster, Fastest
Fill out this page as part of your lab.

LAB TIME

Draw each ✏️ in a box. ✏️ the number 1, 2, or 3 to show fast, faster, and fastest speeds for the 🚗 .

┌─────────────┐   ┌─────────────┐   ┌─────────────┐
│   │   │   │   │   │
│   │   │   │   │   │
│   │   │   │   │   │
│   │   │   │   │   │
│   │   │   │   │   │
│   │   │   │   │   │
└─────────────┘   └─────────────┘   └─────────────┘
      ◯   ◯   ◯

THINK ABOUT IT
Finish this sentence.

_____

- -

We changed the _____ each time.

Which ✏️ helped the 🚗 roll fastest? How do you know?
Try to measure the speed of the 🚗 with a timer. Your teacher will
help you. Tell or ✏️ about what happens.

Walking to Music

PREDICT

What affects how far a person can walk in a set amount of time?

LAB TIME

Work with a 👤 . One 👤 walks. One 👤 records data.

partner

1. Mark a line with tape on the floor. Measure and mark a number at each foot for 24 feet. Your teacher will help you.

2. One 👤 from each team stands on each number. What number is your 👤 standing on?

3. Where do you think your 👤 will be after walking 5 seconds?

_____

- -

_____

4. When the 🎵 starts, your 👤 steps to the beat for 5 seconds.

music

When the 🎵 stops, your 👤 stops. Where is your 👤 standing?

_____

- -

_____

6. Next, your 👤 will walk to 🎵 for 10 seconds. Where is your 👤

standing? _____ Switch with your 👤 and do this again.

THINK ABOUT IT

Would your 👤 go farther if the 🎵 had a faster beat? What if you took smaller steps? Tell or ✏️ about it.

Meter Stick

Use a ![meter stick] to measure how far different objects coast from a ![ramp] .

meter stick

An object's shape controls how it moves over different surfaces.

LAB TIME

1. Use a ![ruler] and ![book] to make a ![ramp]. Line up the "0 cm" mark on a ![ruler] with the bottom of the ![ramp] .

2. Send 3 objects down the ![ramp] . Measure how far each moves using the ![ruler] .

3. Show your answers in the picture below. Choose how you want to number the cm scale on the bottom. Your teacher will help you. Write 1, 2, or 3 along the scale to show where each object stopped.

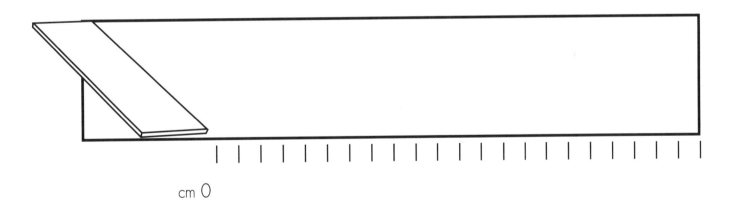

cm 0

THINK ABOUT IT

Which object moved farthest? Why? Tell or ![pencil] about it.

Swing Time

Find out how many you can make in 1 minute!

swings

Work with a 🧍. One 🧍 makes 🏃 on a playground swing.
One 🧍 times and counts the 🏃.

1. Practice counting 🏃. Each swing is 1 time back and forth.

2. The timer says "go" and starts timing. The timer counts 🏃 for 60 seconds and says "stop."

3. Write down how many 🏃 you count every 10 seconds. Switch places and try it again!

The Swing

How do you like to go up in a swing,
Up in the air so blue?
Oh, I do think it the pleasantest thing
Ever a child can do!
Up in the air and over the wall,
Till I can see so wide,
Rivers and trees and cattle and all
Over the countryside–
Till I look down on the garden green,
Down on the roof so brown–
Up in the air I go flying again,
Up in the air and down!

Robert Louis Stevenson

Time (seconds)	Number of 🏃
0	
10	
20	
30	
40	
50	
60	

THINK ABOUT IT

How could you make fewer in 60 seconds? Tell or  about it.

Motion

Matter

Objects have many observable properties, including size, weight, color, temperature, and the ability to react with other substances. . . Objects can be described by the properties of the materials from which they are made . . . those properties can be used to sort a group of objects or materials. . . Materials can exist in different states (solid, liquid, and gas).

National Science Education Standards

Getting Started
To gather information online, enter the query "matter" AND "classroom" AND "learning." Check out books such as **Why Can't You Unscramble an Egg?** (Vicki Cobb, 1990) or **A New True Book: Matter** (Fred Wilkin, 1986), and share the background information below. Check the Equipment Table for materials needed to complete the activities in this unit.

About Matter
- Everything in the world exists as matter in one of four states: solid, liquid, gas, or an electrified gas called plasma.
- All matter takes up space.
- All matter has mass. Mass is the quantity of matter that an object possesses as measured by the force needed to move or stop moving that object. An object's mass never changes.
- Weight measures the pull of gravity on an object. Since gravity changes depending on location, the weight of an object may also change.
- Atoms make up all matter. Atoms come together to form molecules which make up elements such as oxygen or compounds such as sugar.
- Temperature causes a substance to change from one state to another. A substance's mass does not change when it changes states.

Misconception
Share with the students that we know of four states of matter, not three (solid, liquid, gas, and plasma). Plasma exists in the gases of our sun.

Teaching The Unit
Quick Lab Experiment and Data Collection/Communication: Explain to students that everything in the universe, from a rock to a star to a drop of water or a gust of wind is called matter. Discuss the fact that matter generally exists as a solid, liquid, or gas and share examples of each. Ask students for ideas about how to tell one type of matter from another; introduce the activity and discuss predictions. Although students are only exploring solids and liquids in this experiment, include gases in your discussion. Use plastic bottles with lids for samples. Potpourri can be substituted for pine needles. Allow students to pour solids and liquids into different-shaped containers. Be sure students know to write in the letter codes for each sample in the appropriate boxes on the Data Collection page. Students should conclude solids do not change shape, while liquids do change shape to fit whatever contains them. Some solids and liquids will have the same attributes and can be grouped together. For Think About It, students may sort by weight, taste (for approved edible samples only), or ability to allow light to pass through.

Working Together: Classes enjoy making a matter sounds tape for other classes to guess. Allow air time on the tape for class guessing. Running water, coins, paper crunching, and balloons popping are favorites. Link to a unit on the senses.

Using Tools: For this activity, terms "weight" and "mass" are interchangeable. Model using the balance with unifix cubes (nonstandard units). For Think About it, students conclude different solids have different weights or masses. Help students create subtraction problems to compare differences and choose two solids with equal weights to mass.

Learning Links: Instead of object name tags, real items may be placed in the bag (with liquids in plastic), having been collected as homework or on a nature walk. Trade bags and record predictions vs. results. Share which clue was crucial for detection in each case.

A Closer Look at Matter

PREDICT

How are different kinds of matter alike and different?

LAB TIME

Try these four tests with your matter. 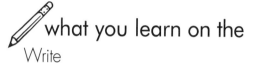what you learn on the next page.
Write

You'll Need:

Water   Rice   Vinegar   Coins   Rocks   Pine Needles   Honey   Shampoo

1. Does your matter change shape? it.
   Pour

2. Does your matter have a smell? it.
   Sniff

3. How does your matter feel? it.
   Touch

4. Is your matter a liquid or solid? Sort it.

THINK ABOUT IT

What other ways can you sort matter? Think of 3 ways to tell if matter samples are alike or different. Tell or about it.

A Closer Look at Matter

Fill out this page as part of your lab.

LAB TIME

Look at the letter code for each matter sample.
Use the letters to show your answers to the questions. ✏ the letters in
the boxes. Some letters can go in more than one box for each question.

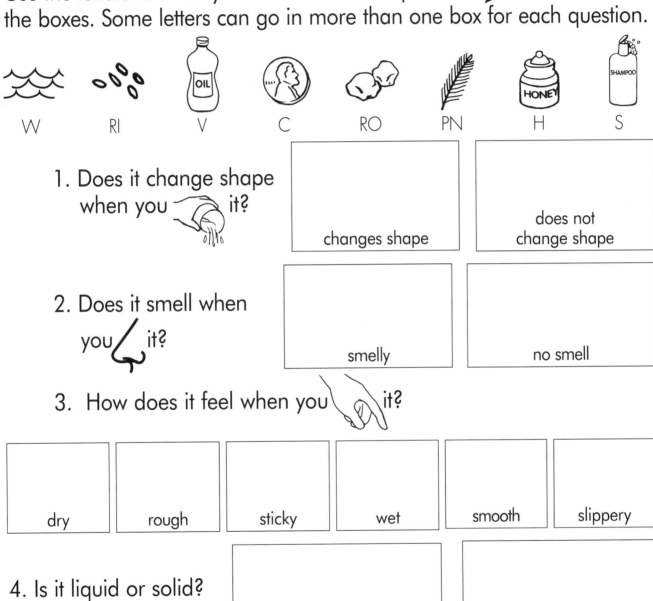

W   RI   V   C   RO   PN   H   S

1. Does it change shape
   when you 👌 it?

| changes shape | does not change shape |

2. Does it smell when
   you 👃 it?

| smelly | no smell |

3. How does it feel when you 👆 it?

| dry | rough | sticky | wet | smooth | slippery |

4. Is it liquid or solid?

| liquid | solid |

THINK ABOUT IT

Liquids/solids do not change shape. (Circle one.)
Liquids/solids always change shape. (Circle one.)

Matter

Mystery Sounds

PREDICT

Can you know a liquid or a solid by the)))it makes?

sound

LAB TIME

Work as a group. Choose 5 solid or liquid objects to [tape].

✏️ or 🖍️ your objects here.

draw

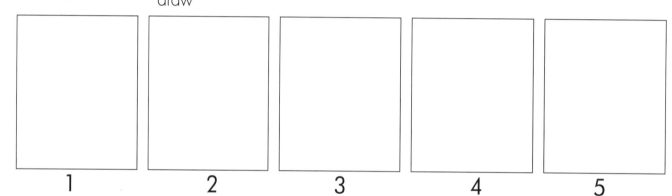

| 1 | 2 | 3 | 4 | 5 |

Choose a person to ask, "What am I?" after each))).

Take turns recording each object's))) on the [tape].

Play the [tape] for another class or group.

THINK ABOUT IT

Finish these sentences.

The _____ was the easiest to guess by its))).

The _____ was the hardest to guess by its))).

Listen to another group's [tape]. How well did you guess each)))?
Tell or ✏️ about it.

Balance Scale and Centimeter Cubes

Use a △ and ◻ to order solid objects by weight.

balance scale   centimeter cubes

The more an object weighs, the more it makes one end of a △ tip. The more an object weighs, the more ◻ it takes to show its weight on the △ .

LAB TIME

1. Choose 5 solids to weigh on the △ . Label your solids 1, 2, 3, 4, and 5.

2. Place each solid on the △ . Stack ◻ until the both sides of the △ are level.

3. ✏ how many ◻ each solid weighs.

   _____ _____ _____ _____ _____
       1   2   3   4   5

4. Order the solids by weight.

   _____ _____ _____ _____ _____
     lightest   heaviest

THINK ABOUT IT

Finish these sentences.

Different solids may have different_____.

The _____ weighed _____ more cubes than the_____ .

Use the △ to find two solids that have the same weight.

Tell or ✏ about it.

Matter Detective

1. Think of a solid, liquid or gas.
   ✂ cut out and fill out the ▭ .
   name tag

2. Place the ▭ in a 🛍 .
   paper bag

3. Fill out the 5 clues. Do not fill
   out the answer.

4. ✂ and 🧴glue all the clues and the blank answer to your 🛍 .

5. Trade your 🛍 with a friend!

Secret Matter Name Tag

– – – – – – – – – – –

(circle one)

solid   liquid   gas

Secret Matter Clues

1. I have a ----------------------------- color.
   _____

2. I feel ------------------------- when you 👉 me.
   _____

3. I smell------------------------- when you 👃 me.
   _____

4. People use me to -----------------------
   _____

5. I do/don't change shape when you ✊ me.
   _____

   ANSWER: I am -----------------------
   _____

THINK ABOUT IT

Which clue was most important for guessing?
Why do you think so? Tell or ✏ about it.

Matter